ADVENTURES WITH MY HORSE

Also by Penelope Shuttle:

The Orchard Upstairs (1980)
The Child-Stealer (1983)
The Lion from Rio (1986)

PENELOPE SHUTTLE

Adventures with my Horse

Oxford   New York

OXFORD UNIVERSITY PRESS

1988

Oxford University Press, Walton Street, Oxford OX2 6DP
Oxford New York Toronto
Delhi Bombay Calcutta Madras Karachi
Petaling Jaya Singapore Hong Kong Tokyo
Nairobi Dar es Salaam Cape Town
Melbourne Auckland
and associated companies in
Berlin Ibadan

Oxford is a trade mark of Oxford University Press

British Library Cataloguing in Publication Data
Shuttle, Penelope, 1947–
Adventures with my horse.
I. Title
821'.914
ISBN 0-19-282218-7

Library of Congress Cataloging in Publication Data
Shuttle, Penelope, 1947–
Adventures with my horse / Penelope Shuttle.
p. cm.
I. Title.
823'.914—dc19 PR6069.H8A65 1988 88-4857
ISBN 0-19-282218-7

Typeset by Wyvern Typesetting Ltd
Printed in Great Britain by
J. W. Arrowsmith Ltd., Bristol

For Peter and Zoe

Acknowledgements

Grateful acknowledgements are due to the following magazines and anthologies in which some of the poems have already appeared: *Outposts, Words International, The Manhattan Review, The Poetry Book Society Anthology of New Poems 1987, Poems for Shakespeare 1987, PEN New Poetry II, Poetry Review, Stand, Falmouth 1972–1987 Poetry Group—a bouquet for Derek Power.*

Contents

Jungian Cows

In Switzerland, the people call their cows
Venus, Eve, Salome, or Fraulein Alberta,
beautiful names
to yodel across the pastures at Bollingen.

If the woman is busy with child or book,
the farmer wears his wife's skirt
to milk the most sensitive cows.

When the electric milking-machine arrives,
the stalled cows rebel and sulk
for the woman's impatient skilful fingers
on their blowzy tough rosy udders,
will not give their milk;

so the man who works the machine
dons cotton skirt, all floral delicate flounces
to hide his denim overalls and big old muddy boots,
he fastens the cool soft folds carefully,
wraps his head in his sweetheart's sunday-best fringed scarf,
and walks smelling feminine and shy among the cows,

till the milk spurts, hot, slippery and steamy
into the churns,
Venus, Salome, Eve, and Fraulein Alberta,
lowing, half-asleep,
accepting the disguised man as an echo of the woman,
their breath smelling of green, of milk's sweet traditional climax.

Killiow Pigs

(From Killiow Country Park, near Truro)

Five adolescent suckling pigs
fanned out alongside their sleeping mamma;
each daughter big as an alsatian dog,
her five petticoat-pink starch-skinned girls.
They sleep with resolution and vitality.
Our admiration does not wake them.
Fed on apples, their flesh is ready-seasoned.
This afternoon heap of pig breathes a clean dusk
into the air; spring and dung,
rhododendrons, sour vapour of swill and straw.
With their sexy squiggle tails,
their ears soft as cats but big and lopped-over
like ambitious rabbits, with their long carefree
strokeable backs, their feet comic and smooth,
snouts succulent,
these sisters lie outspread, five cordial orchids
against mother's blushing pungent bulk,
dreaming of orchards
where an exiled male roots and roams,
his boar thighs tough and angelic,
his head lowered to the cool brisk echoes of morning,
his ringed nose a gleam of gravity,
his sudden stillness all swinish magnetism.
Dossing mother and daughters quiver in sleep,
the juice of desire lolloping over their lips;
snouts swell with love; tails uncurl, grow fine
and tender as silk;
each meets her orchard lover,
dreamy pigs in their matrilineal slumber.

As we watch these females, hope and desire
rise in us, a cloud of matrimonial heat,
blossoming and getting the better of us,
oh these shameless porcine arrangers of marriages!

Alice

I live in one room.
My bedroom is my kitchen,
my study is my bathroom.
I am absorbed by my own powers,
feeling beautiful and resourceful.
I am awaiting an avalanche of young.
In me fifteen new hearts beat.
My stretched belly-skin is near splitting,
my bulk is pastoral, I know.
I smell of melons and cheese.
I am not restless or nervous.
I look pityingly at you
who don't possess my one room.
Soon there will be such a squall of piglet,
a shoal of tails and tingling ears,
an april fall of flesh,
a sixty-legged blind creature,
not a scratch on it;
a chute of pig, shriller than puppies,
fitter than fleas. I know all this
from previous experience.
Every one of my imminent litter will possess
our breed's gift for caricature.
The dog will turn and run from their chaos.
They will not be dangerous in their cherry-pale
and sugar-bright skins; but loud.
In the paddock they will race and scamper.
Like mine, their lives will be immensely public.
Under the afternoon sky they will sleep
as babyishly as in any cartoon
that would bonnet and bib them,
as if their flesh were lifelong safe, inedible,
and myself, Alice, their mother,
a human mother resting with full breasts bared
and aching in the flickering shade of the mimosa tree.

Alice's Husband

He is both predictable and unpredictable.
Both gracious and fierce, heroic
and brutal, a gothic husband.
Just when I have forgotten him, he returns,
savage and hopeful, pale both with lust
and with an aesthete's melancholy. This
was not necessarily his idea, he means. Elvish,
his turn of head; his glance by turns
high-souled, gloomy, flirtatious, deferential.
Glowering, he sniffs towards me. Ringed,
his snout is merely embellished, not owned.
He draws my smell into his nostrils
with a shudder of scandalized disbelief,
then trots a little closer; all this
surveillance! He moves forward, he who
has no after-life. His back is long,
his thighs huggable, his tail a perfect solo,
his odour of honey, pepper, camomile and wax,
a reek of desire, my infinite temptation, his bait,
this fallen angel of the pigs.
His testes are charming, burly,
they billow out, ballooning
from sudden excess of emotion; his plum-coloured
tongue lolls.
His powerful torso is like the beginning of flesh,
his massive capable jaws good as any dog's;
distract him from me now
and he'll charge you, speed and force, for the kill.
He is the axe at my root. His weight on me
is aficionado, apostle, family-man,
a giant refreshed.

Dressing the Child

Wide spindling river,
out-of-fashion green and cream
woollen river cast on huge brown needles,
bare tree-trunks garter-stitching
the currents, purling the banks,
armholing the bridges, casting-off barges.
The winter river is a long coarse sleeve
of flecked and brindled wool
into which I shove your thundery arm, the cuff
of ribbed bushes scratchy-tight on your wrist.

God Dividing Light from Darkness

(*Michelangelo*)

An old man,
a feeble agèd grumpy
grimacing old god in the clouds,
cloud-bearded, lear-haired,
clambering through womb-mist,
fumbling, lost, tunnelling
through epiphanic coils,
their foams and jades and grimes,
their hulled and frilled scapes.

As old as these nesting clouds
that water-lily the void together,
he throws up his fog-robed arm
and delivers the world;
out of the womb-age of the old god
comes a cloud of earth and sea,
sky and dream,
ripple of sand and ripeness of rock,
fissure, passage and cave,
leaf and coriander seed,
swamp and hedgehog, etcetera. . .

He bears this child of his old age
without tenderness,
enraged, his heart closing
like a book of stone,
his eyes changing colour each instant,
flickering kaleidoscopic lightning;
his robes radiant waves
of cold-rainbows,
this frowning merciless old father,
minus cherubim,
bristling with opposites, the pangs of creation,
the light his perfect son,
the dark already too bad to be named,
serpent turning and grinning,
prodigal on the balance-scale.

Snakes and Quakes

Anything that wriggles
might be a snake.

Our yard grass whispers
serpentine, stern; seeding
into neighbours' neat gardens.

Coil of cloud returning
with rattling-tail;

the rain a green viper
withdrawing from life,
reclusive in the water table.

Around the new moon,
a snake of light
shakes and

the air cracks its whip;
window glass shatters;
buildings shake and boom.

The snake opens his eye.

Houses, ready to fall,
exhale the scent of him,
that breath of earth.

I sip water,
fearing everything but the snake.

I hold him,
spin in woman circles with him;
his hard dry lip against mine,
his tongue mustard,
colour and taste.

He sips my breath.

Doors move suddenly
with no help from hands;
the sky galloping,
rivers zigzagging,
the air's soft growl rippling.

Plumes of trees twist,
green and baffled.

But the serpent is already back
in his secret corner of thin air,
long-necked and unanswerable.

The earth locks its stable doors again.
The horse of quake gone.

I, snake-collared, yawning, garlanded
and sashed with serpents
invisible but coiling and scolloping
electric eelishness around me,
watch the golden worm of lightning (my beloved!)
fork his brazen tongue over the black sky,
lisping flame, stammering fire.

Thief

He will steal it, whatever you possess.
Whatever you value, what bears your name,
everything you call, 'mine', he will steal.
Everything you have is frail and will be stolen from you.
Not just watch or bracelet, ring or coat,
bright objects, soft splendours, gifts, necessities,
but the joy that bends you easily and makes you feel safe,
your love of what you see each different morning
through your window, the ordinary seen as heavenly.
Your child's power, your lover's touch, will be stolen
from under your nose. He will steal everything.
He will take everything from you. You will never see him.
You will never hear him. You will never smell him.
But he will destroy you.
No surveillance is close enough, no guard clever enough,
no lock secure enough, no luck good enough;
the thief is there and gone before you have sense
of breath to cry out.
He has robbed you before, a hundred times.
You have never seen him but you know him.
You know his vermin smell without smelling him,
you know his smile of learning without seeing it,
you feel his shadow like deprival weather, grey, oppressive.
You know he watches from far away or from just round the corner
as you regather your little hoard of riches, your modest share
of the world, he watches as you build your shelter of life,
your hands raw from working day and night, a house
built out of bricks that must be guessed at, groped for,
loved, wept into being; and then upon those walls
you and your people raise a roof of joy and pain, and you live
in your house with all your ordinary treasures,
your pots and pans, your weaned child, your cat and caged bird,
your soft bestiary hours of love,
your books opening on fiery pages, your nights full
with dreams of a road leading to the red horses of Egypt,
of the forest like a perfumed pampered room wet with solitude.

You forget the thief. You forget his vanity,
his sips and spoonfuls of greed. But he watches you,
sly in the vaults of his wealth.
Shameless, sleepless, he watches you.
Grinning, he admires your sense of safety.
He loves all that you love.
Then, in disguise, with empty pockets, his fingers dirty
and bare, rings of white skin in place of gold bands,
he comes like a pauper on a dark patchwork morning
when summer is turning round and robs you blind.
He takes everything.
He is the thief in whose gossamer trap you have been floating
all these years. He comes and takes everything.
Your house is empty and means nothing, the roof falls in
and the walls of love dissolve, made of ice;
the windows no longer watch out over heaven, the bare wooden
floors show their scars again and ache for the forest.
He takes everything you have, this thief, but gives you one gift.

Each morning you open your eyes jealous as hunger, you walk
serpent-necked and dwarf-leggèd in the thief's distorting mirrors,
you go nakedly through the skyless moonless gardens and pagodas
of envy that he gives you, the thief's gift, your seeding wilderness.

Virgin Martyr

(St Barbara, 7th-century Christian martyr,
who in legend was imprisoned in a tower,
tortured, and murdered by her pagan father.)

His fisted hand
in her tangled hair,
he drags his daughter,
the naked virgin,
up the spiral stair
of a stone tower
beside a rushing river;
deaf to his roar,
his curse,
she kneels to pray
in the nakedness
of her bud breast,
bruised knee and racked heel;
in the stone room
at the top of the tower
her father tortures her
with whips and hot irons,
breaks her fingers and toes
one by one,
urging her
to worship his pitiless gods
but she will not
so thrusting his hand
into her matted braids again
he bares her unresisting throat
to his knife
jerking her head back
and slitting her flesh
with one unswerving stroke
as if she were a young deer
and her blood flows
from the top of the tower

into the rushing river
reddening it
and lightning with its fork of fire
strikes the father in the tower
and the tower falls down;

she is at home
in the flames of Paradise,
her throat a radiance,
her fingers and toes good as new,
her resurrection into after-flesh
serene and ecstatic;

she soon finds
her fearful father
howling like a child over his own death
in the dark of hell,
crying for the mother of souls to pick him up
and suckle him into quiet.

Overnight

I am pinning wet laundry on the line an hour
before summer bedtime,
my shadow in the moonshed night holds up
warm dripping gathers, soapy scallop-edged hems,
a smocking of wrung-out blouses
shepherdessing me through my chosen task.
My rough fingers pin up his hugging shirts,
his ten long white arms fly out in spirals of spray,
the wet tails of my skirts twist to snares and nooses,
a daughter's near-adolescent fashions tremble and dance,
the bath towels pull like tarpaulins. I haul
them all up on the line, giving them to the moon
for drying.

I thought I was not meant to be loved,
but these wet clothes weigh me down with love,
its luscious clumsiness, its terror and wit.
I look up at the moon.
He will do his share of the work, I know,
even though he's only at half-strength;
all night he will dry the clothes with his clean clannish breath.
They will not be called strangers by him.
I leave him bending forward over the garments like a lover
and come to you.

We are beyond clothes;
naked, our bodies pillowed and spelling out breath,
we are the long kite-ribboning lovers;
my flame of orgasm is innocence returning, yours breaks
on me like a sky of connubial indoor rain.

Out in the yard, the washing sways and lulls;
solemn as children, the pale plump dresses,
the collars and cuffs with their couvades of lace,
the fledgling buttons flouncing in moonlight,
a row of fluttering sentries not needing their colours

until morning,
when I slouch sleepily out and unpeg them, creased
and armour-stiff, missing the moon, fearing bird-breath,
eggshell-bridge sky, the expectations of day;
in the house they lie like elderly rustics,
awaiting the phoenix of the iron to smooth them back to life.
Alive again, they desire and cover our family nakedness.

The Old Man

Often after we make love
I dream of your agèd father
as if our lovemaking
has led back to your beginning,
your sire roused by our pleasure.
In the vigour of his age
he's like an old charioteer
telling us that what we ride on
is our life,
what we use as our vehicle
is the breath
one day we will let go of forever
as for an instant
we let go of it in love.

Zoe's Horse

White morning of cage-birds with their voce di testa.
I fall diamonded into second place.
The bed's earliness shines as I shake and shush the towering quilt.

Under the wrought-iron garden seat of the old school
there's a child-made nest for outdoor birds,
there's velvet thunder, a flower
thrown into its own arms.
From different windows, I watch colours
not meant to be named, not meant to be loved.

Bees in the tearful bush by the unlatched gate
brush secludedly against its syrups
as the child walks languidly by, dragging her torn nets
of translation, her paper kite, her collected petals;
she is singing about her subjective white horse.
What has she hidden under the upturned enamel bowl
that shines out of tune on the path
between softly-rouged and sound-sleeping flowers?
Mouse? Peach stone? A ribbon to rein her horse?
She will not say, not even when heavenly rain falls on us again.
She has the indifference of the coquette, the shrugged shoulder.

It's part, I guess, of this gamble of ours,
that we share, to be perfect as the rounded moon.
(Mothering and daughtering moon!)
Stretching, to be perfect.
She, to the future.
I, with one long stiff step
landing in the 'then' I still call 'now'.

Zoe's white sung horse is watching her.
He kicks and prances. He is her care,
her trouble, her love.
His eyelid, lip, flank, and mane
are white as a wedding. His tail foams

like mother's milk, his hooves are windfalls.
Heartless horse, he gallops off, wide-eyed,
his breath pure as chance,
her lunarian, running, racing, white as the bosom,
collar and cuffs of a fine lawn shirt,
white as napkins shaken out.

'Oh he's gone,'
she says boastingly.

Summer mists hide him
like a pearl in a pond.

Zoe in cool skirts waving,
making no wishes.

The Sugar Lump

A simple horse,
(he's a child's horse),
doesn't have to love you
to get his meals;
because his white-lily-and-lotus mane,
his rough-rouged hooves,
his copper and tortoiseshell glance deserve such sweetness,
and when you ride him
you taste that same selfless sweetness
in your mouth.

Adventures with my Horse

1

If the boy is a horse, who rides him?
He is a boy when he fears the night.
He is a horse when morning comes.
Who will ride him in all his flying colours?

She is baffled and careful in blue,
watching her child gallop long daytimes.
If he is a horse, can he understand November?
Can he do desert arithmetic? Can he count grains of sand?

If the boy is a horse, what can she promise him?
Who is racing him past the silence of greenhouses?
Whose posse does he join, then suddenly leave,
wheeling into shadow and self, untiring?

The woman fears him.
He races from her, unteachable.
He cannot learn to be a baby again.
He is capable of neighing.

His rough-maned head glows in windswept sunlight.
His limbs anger her with their faithless strength.
Around him she sees the hovering battle-dead,
boys who were horses.

2

If the boy is a horse, why does he write his name in invisible ink?
If he is a horse, why is he not afraid of fire?
Why does he wake up yelling his wills and won'ts?
If he is a horse, why does he want her to play the piano for him?

If the boy is a horse, why does he tell lies?
If he is a horse, why does he watch scornfully
as his mother scatters corn to her pretties?
If he is a horse, why does he swim in the forbidden weed-clotted
    river?

If the boy is a horse, why does he love his mother's brooding
    cupids,
her armoured angels, the yawningly-safe brocades of her room?
Why does he stroke her broad-sashed green kimono?
If the boy is a horse, why is his sex shy but dizzy-sharp as a needle?

If the boy is a horse, why does his mother read him stories?
If the boy is a horse, why does he close his eyes,
lie a dead weight in her arms, stung by grief,
his eyelashes blurred with tears busy as bees?

3
The woman is alone with her boy.
He kneels between her knees, secretive, childish, wary.
He has the big dubious eyes of a foal.
His heart rides towards her.

She looks at the boy she never meant to love.

At last a mother's passion rises in her,
intense as sex, demure as a colour she cannot name.
At last maternal love settles over the woman and her son,
its blazing quilt stabling them.

Now the boy is just a boy, lazy, tousled,
heroic, flirtatious, her dandy, her April Fool, her son.
Her breath aches in wonder, as if he had just come to life
and the blood still boiled from her.

Her breasts throb for his greed.
Her arms reach for him, then let him go where he wants.
The boy colours his picture of a horse leaping.
When he hides from her it is because he is her favourite now.

Five Carp, Two Swans

The big heads of five saffron carp rise from the pond,
indecent-lipped, impressionistic and impish,
their scaled shoulders fond of the weight of water,
coarsely-glimmering in the sandy six-inch shallows;
their slothful aunt-like eyes watch our blue shadow hands
stretched out to them in as deep a languor as theirs;
like sucklings they slubber up cake from our fingers,
their lips delicately-snarling, pinkly-raddled;
they graze on bread from my daughter's flat cautious palm
without tickling, tender as ponies.
They are like tasting souls, lifting gourmet heads above water,
testing nature's heart on this side, air too bright,
but food so fine.
We could feed them anything, plums, dog biscuits, bananas,
but in glinting unison, a hand of fishes, they shoal and slither
to the darker centre of the rippled pool,
as one black honeymooning swan females towards us,
her eye vain, cool and greedy, jostling air with her lipstick-red beak;
we poke the last of our picnic crossly into its dry cochineal clack.

With fin de siècle certainty she rejoins her watching mate,
fossil-black wings folded, grey paddle-feet idling;
the swans share silence, their marriage is mute,
their long-necked love is made of circling,
of drifting towards night, of being black,
of going grandly on the water, fish belching beneath their feet.

The Living

The dead man's last breath is soft and limpid,
egg without shell, sigh natural as rain or moon;
he grips the chair arm, it is web or spray blown over water.
He falls beyond the terrors of oxygen
into a perfume of ghost, sour pollen, sweet cumin.

The cat's deifying eye, grown soldierly in the dawn,
is watching him die with the calm curiosity of an immortal,
animal kneading long necessary claws on a worn solitary mat.

The dead man was an adder-up of happiness,
an advocate of memory, its glitter and scope;
fallen now from an old chair's indoor velvet,
he smiles, half-pleased in a crouch of farewell
on the bright fictitious summer of the floor,
his chance-blue shirt becoming an airman's
camouflage to hide him from the sky and its blue futures.

His cat goes unflinchingly through the doorway.
The room is empty of breath and thrown into its own arms.

Light comes in restless from the garden,
looking for its master, who is silent,
his workmanship over,
gone on his solo flight.
Female and blind, the windows ache against no one's stare.

Outside, on the house step, cat licks his unsoothed fur,
tongue rough, impatient and rational.
He tries to be good, but see how anger
ripples his spider-black coat,
how he stalks his prey with an executioner's reluctance
and pride,
deaf to the sky-high voice from the house
plunging into a lake of grief, of love and duty;

the woman's living voice demanding,
is the old man already as old as this?
and remembering into tears how he has begun to haunt her.

2

Years later,
dreaming that the long-dead man has died again,
she wakes to all the white news of grief
as if for the first time;
this bare table, that open door,
bridges, streets, hills white as porcelain,
the white kitten racing
through a vulnerability of rooms, breakable kitchen,
trembling bathroom,
white sky dropping to white ground,
each second long as a change of moon;
and the white abyss of stone under which he lies
presses down on her, its tons of sky,
till only the white print of his heart remains,
no flower opening all the white summer,
white lawns stiff with the daylight of the living.

Her Butterfly Husband

She re-enters earth's atmosphere,
pulling the sky in through the window
after her;
her bed is blessed by her dream of having
as many blue dresses as her arms can hold.

Her butterfly husband wakes.
His shadow falling on her is cockled,
creased and ridged. Delicate honeymooner,
he greets her jealously, gliding away,
his own blue better (he means) than her dreamed blue!

She strokes his heels, saying
she slept like a fool. Behind shutters white as egg,
behind doors curtained with the cool of white,
she coaxes her butterfly husband
back to her unpatterned sheets, her palais de vérité,
her affectionate ceiling; she charms him
with her implications of luxury,
her coddling. He comes to her slow as a listener of mandarin.
She kisses him, her tongue
a long handle of light around his virginal sex.

On her mirroring pillows, in sweet and lustful recumbrance
he rises on his first flight,
her butterfly husband is rising and fluttering,
caught fast by her petalled ovarian smell,
his deflowering her easy and pleasing chore,
his sigh coinciding with hers, a sky circling them.

Next she brings him a married breakfast,
turtle soup and sticks of cinnamon,
water and honey in a white cup for her unbridled flying husband.

Passion

He draws memory out of me with hands of fire.
His touch is miracle and shock.

My masseur strokes me into swallow-light,
torrents of dove.

He kneads me. His fingers are wildfire and blush,
rose and scorch. I bask and singe. His hands burn me.

I moan. He firms the fiery earth of me.
He sears the air of me. My flesh furnaces.
(But his kiss on my throat pulses a cool radish scent.)

We loll and sweat on his plush throne, on far-flung cushions.
I tremble under the tact of his touch.

His tongue floods me like honey and cirrus.
I slide into my own sheet-of-fire ghost.

2

Like a god humming and making things
he puts his lips against mine. Oh I shiver!
I freeze! He is ice. He is knife-bladed and bleak-beaked lover.

Now I am his oblong silk scarf left to chill in the freezer,
along with the icecream and the raspberries stelled with frost.

He is frost-bite.
He kisses each nipple to zero.
His sex is a prism of ice in me.

Now the sky calls out for its birds in my voice.
Now a door of dawn slams.

Our reign of henna and sauna is over.
He flies to heaven on my warm heart.

3

When I wake
I turn sweetly in my blanketted bed,
(that wanderer haughty in his own home).

My wrist-falling white ruffles
flap and float till I'm naked as a horse.

I wrap the dress-length silence around me,
hands skimming my waist.

When our baby under his broidered awning
opens his eyes to a breath of birds,
I can believe in the kindness of hell,
the dark ages of heaven.

4

If my devout and carnal brother,
my lover of fire and ice ever
steps into daylight,

I will baste that satan with sunshine,
he will fall from my lips,
as I fell from his.

Her Fox

Marble fox with eyes of fire,
haze of rain on chill fur.

His snarl drips,
his brush is tracked with frost.

Clouds hunt him, a dark pack,
but he keeps so still he is safe.

He is indissoluble,
white as a child of the lightning.

His wide-held jaw is only traditionally savage.
He is no sabateur.

Let me, he says,
rest my head in a woman's lap,

she who is both fruit and furniture,
she will nurture me,

my stone-white heart is the natural partner
for her tether of days,

I am her silent animal frozen past fear,
her garden companion,

her lap's slight camber warms my stone pelt
to a semblance of flowers,

and her sharp-as-knives tears
blood me with their salt, their calm.

Draco, the Dreaming Snake

Like a sigh in silence, the serpent,
oh, he dreams me, in his solar, in his naked house.

The serpent weeps if he thinks he is not wanted.
But he is. He dreams me. In his solar.
In his room for sun. In his pond of light.

Oh his great energy and his cleverness,
out of the naked house of the virgin he creeps,
skill and shame mixed in him,
and a vapour about his skin I shiver for,

glimpsing how it works, his magic,
but only a glimpse, he is all his own magician,
he sleeps as he travels over the clays and granites

of the land, never unemployed, always a worker,
yes, he gathers up the whirlwind of the sleepless
and finds some sleep for them. That is his work.
He dreams me. My serpent dreams me.

It is his intimidating but helpless gift.
In a fountain of pure undwindling water he dreams me,
he dreams me in the soft male heart of the lily,
its innocent roar.

He dreams me in my dream of the paper dead, the lost parents.
He dreams the shadow that does the washing-up
and the ironing in the new house. He dreams Mother Lion.

He dreams he is a serpent
who will be the last one of us all to die.
He dreams that all he wishes for will come true.
He dreams me.

In my hand I hold a golden coin covered with blood,
He dreams this.
He dreams this baby under the blue flower of my heart.

He dreams the fragmental stealth of my spirit.
He dreams my future, he dreams my past.
He dreams the breath of this bare room,
the chimney's old ache of blackened brick,
the ceiling a caul of faded paint,
the walls objecting to windows on principle,
doors opening and closing in an ardent future,
causing horror, fear, delight,
and all these dreams move in me like sex,
with little or no punishment or revenge.

Such is the serpent's business,
making something already beautiful even more possible,
the hiss of his hope touching a nerve.

Ancient serpenting tourist, he travels the world,
naked as celsius, naked as the great sophistication of glass,
naked as the flare of spice from your aroused skin,
naked as whatever again and again stretches its long coiled self
towards us, dreaming us and asking for our dreams.

We dream him a new skin to cocoon his aspen heart,
to clothe his whip of a spine; a perfect fit;
tender as an eyelid;
his new scales glitter like rain on his hooded head,
and he dreams us, dreaming him.

The Horse Who Loves Me

The horse who loves me is strong and unsaddled.
He desires to learn nothing.
He sleeps standing, like a tree.
He lifts dawn on his willing shoulders.

I ride the horse who loves me,
hands twined in his bashful mane,
knees gripping his nut-butter flanks.

The horse who loves me goes on tiptoe,
his hooves tap the fiery earth.
The long leisure of his muzzle pleases me.

His smell is salt and primroses, honeycomb
and furnace. Oh the sweat of his glittering tail!
How he prances studiously, the horse who loves me.

The horse who loves me has no hobby but patience.
He brings me the gift of his honesty.
His big heart beats with love.
Sometimes he openly seeks a wife. But he returns to me.

The horse who loves me is one of the poor of Paradise.
He enjoys Paradise as such a loving horse might,
quietly watching the seven wonders of the night.

2

Look at my horse!
His neatly-plaited snowy tail
hangs like a fine finger between his pearly buttocks.
His name can be Desire, or Brother.

He does not complain of my weight on his back
any more than darkness complains of its loneliness.

The horse who loves me
wields the prick of pain that caps the dart of love.
We gasp at its pang,
then race for the scaled wall of the sky.

The horse who loves me
takes me beyond the lengths grief goes to,
beyond the strides joy makes,
beyond the moon and his sister the future.

This heaven-kissing horse of mine
takes me with him to his aerial home.
Below us, roofs grey, fields fade, rivers shiver, pardoned.
I am never coming back.

Lovers in a Picture

On a bed like an intimate stage
the lovers embrace between red curtains
caught on five gold rings;
the soles of her feet
and the tips of her toes
are scarlet as some phoenix
her red fingertips have held;
across her face turned from him
is the faintest veil;
otherwise she is like him
naked to the waist,
then swirled in big clinging pants
of crimson silk;
his face as smooth and passionate
a profile as she
on their red-curtained Indian couch,
like sonneteers on a rose-patterned mattress;
the two pearls hung in his pierced ear
quiver and her long looping strands
of pearls that fall from neck to waist
and meet behind her back in a shining halter
shiver with a similar suspense;
familiar to us, his leaning towards her,
his concentration and hope;
familiar to us, her mouth,
her small round kind breast;
familiar to us, her knees he kneels between,
familiar to us, his heart-beat, her breath;
they wait in stillness
for us to see how their watchful ease
between the curtains,
their preliminaries and his hand
beneath her elbow
mirror the only way of solving
the redness of those curtains,
the treasure of pearls,

of feeling the air lifted up
on its golden rings
and rocking us;

familiar to us, these lovers
at their work of guidance and love;

and night's kohl drawn across our own eyelids.

Your Girl

Not knowing you are a horse, your girl gives you a glass of wine.
Not knowing you are a horse, she flies through your gloom
on a cloud of family flowers.
Your girl takes your life in her hands,
lighter than a bird of glass her breath trustingly manufactures.
She has not fallen in love with a horse.
She comes to you because you are frail and human.
You sleep with her in a low and lacquered chinese bed,
on silk-fringed pillows of peony and sunflower,
under a water-lilied and willow-treed quilt of satin.
On the wall, an intimate embroidery
of Aphrodite contemplating her youngest child,
drowsy and triumphant on his mother's couch. . .
(You embrace under the oriental covers,
the lamp calmly-shawled in its stiff-eyeletted shade of white card
throwing a ruff of shadow on her face. . .)

The night knows you are a horse
but your girl does not, despite her wisdom, her honesty,
her unselfish and amorous mouth, her skin sheer as white poppies.
What a fabric she is for stitching, you are the thread
moving in her, your tears hidden from her.
Such a sigh you draw from her, like water from the river in Egypt,
Nile flood, heaven caught in the net of her sex,
her own smell sudden rain on leaves and brambles,
or the blue necks of doves.

Then she sleeps in the dazed bed.
She sleeps in your arms, your girl, your bedouinne.
She dreams you are a horse.
She dreams she forgives you
and that she canters through favourite rooms on your strong wide
    back.

Mother and Son

No longer as happy as you used to be,
you guess your son is dreaming of prize-fighters,
of their polished black gloves, leather-fists
monstrously-swelled, and the dry sour fit of the gumshield.
His father is one of the ghosts but the boy is not frightened.
Around you he still creates harem music's boldness,
a hanging bed in a large and luminous room.
The boy is your nestling, your prince of pearls.
You feed him peppermint swans and sugar pelicans,
hoping he'll stay with you.
But he's grown, and his father's look strengthens
stroke by stroke on the boy's face,
a pencilled mask of a man.
His father offers threads of truth,
sweet hairy cobwebs of masculine virtue.
The lost father calls him, low soft coaxing commandments.
After a last kiss,
your boy moves from you in a contented cloud
of serious considerations.
He's lost his hammock sleeper's grin.
He scuffs up the insolent white dust
of the road his father made,
stamping down its flat even stones,
his chosen route leading him along avenues straight as rulers,
between ditches burnt to a swarthy bristle,
reeking of butane;
he walks crescents of houses that curve windowed and sober,
he comes to broad-backed well-shaded squares,

bringing his question
into each cluttered courtyard of strangers,
all his father's former addresses,
all the searched districts of poignance,
the boy trekking further from his mother each day,
forgetting her midsummer labours,
learning to live with men.

The mother moves through her double-doored days
with the sweet flinch of blindness.

His letters come like birds homing.

Like a hindu labyrinth, her heart holds its tears forever.

Seventh Son

1

I brand my six sons with name and nickname,
then I forget them.
They grow swiftly to men.

Each goes his separate way,
restless, clean-shaven or licentious,
hunch-backed, optimistic or snake-hearted.

One plans to avenge lyricism.
One puts the beauty of his breath into a horn.
One will balance on the high wire,
the next spread riddles and rumours.

The fifth will sail the world,
the sixth perhaps turn magician.
None will ever hang again from my nipple,
like fruit of a tall sweet tree.

2

My seventh son waits to be born.
Night fathered him, riding me.
Nothing can stop him.
He grows in me calm as any root in nature.
His hands open and close with his own artistry.
His eyes open wide, change colour, flicker like skies.
Already he has the gifts of his siring;
strength and solitude. He belongeth.
He does not fear his lack of brothers.
He is to be my firstborn. He has promised.
My body is light with his weight, his footfall, his life.
He waits for life like a blossom-coloured horse.
He is fit for pastures, my true son light as a dove in me.
I long for him, his boy's games, his cocksure debates,

38

his exacting demands. My paloma in the dark, my hard-hoofed son,
I carry you easily as the tongue in my mouth.
We race over the pampas, we are one creature, unsaddled, free.

You break out of me softly as rain the light adopts,
you spin your strong bones out of me in one great bloodied ache of
    comfort,
I give one shout of recognition,
you lie kicking against my sweating flanks,
kicking, my colt, my wild-maned one;
the boy is breathing,
not yet angry; still half horse. . .

The Summer Visit

Today the most difficult ghost returns,
smelling feminine, looking casual,
wanting everything.
She walks into my house without a word.
She watches how I do things,
polishing, mending,
studies how I breathe,
how I lift the silent white cup
from its hook,
or lean sorrowfully out of the greatly-loved window.
She smiles as I move a chair
into another room, for the sake
of a child or a cat.
She greets each sigh and ache
with a nod of severity.
Her chill seeps into me,
till my breath
is a white loop of ice,
my skin an armour of shiver,
my sex a frozen crack.
Her white hand on my heart
is beating like rain on the garden roses,
her silence goes over my head
like a steep and stealthy wave,
I gasp and weep,
she looks at me with love.
Her kiss is brine, plum, aloe, honey.
My mouth opens, haunted.

At evening the live rain begins again,
driving the ghost back to her cold alcove,
washing away her power to echo and enchant.

After she's gone, the summer trembles giftedly in my hands,
then stills, lying warm as an egg from grandfather's chickens.

The Fig Tree

The day cometh.
Last night the moon's brother flew from heaven.
He brings pleasure to all his beloveds.
He loved us both. He touched us both.
He left you sleeping like a boy-bridegroom,
helpless, serious, gold tear of sleep
flattering your lash.
He sent me dreamless between the spate of sheets.

All through the summer night
our lamp stares from its low ruminant table;
our forgotten vesuvian.
A fiery morning wipes out lamplight.

The sun, inquisitive, acquisitive, dextrous,
makes such gestures as I expected,
his luscious tongue of gold can't hold back
from licking everything,
the smack and slurp of him scuds over my face,
is he dragon or dog? I lift my hands
like flimsy rivals. You sleep on.

I slip through the faded curtains,
step on to the eloper's balcony of rusty curlicued ironwork,
the metal warm already to my hands.

On the fig tree, that nursing mother
squat and heavy in our neighbour's run-wild garden,
the burdened branches are massed
with big Adam and Eve leaves, sturdy and decorous,
and the coarse bottle-green figs clump and cluster,
solid, swollen, fed on willing milk.
Overnight, they have ripened; as we slept
they came to the understanding of all their hope,
and now hang ready for hands or a breeze to catch them.

41

I look and bite my tongue in pleasure.
As I lean from the balcony,
forgetting even you, they stare me in the face,
graver than plums, greener than apples,
each one an exact partner to my hungry heart,
my tasting lips.

I breathe in a sweetness deeper than honey or clover,
wiser than mint, purer than water.
I can taste it to the shock of the kernel.

The day cometh.

CLAYMAN, LEATHERMAN, AND GLASSMAN

CLAYMAN, LEATHERMAN, AND GLASSMAN

The Eves are coming. The Eves are dancing
and working, whispering and flattering, deceiving
and observing, crouching and licking, stretching
and bargaining; the Eves are coming,
a hundred, a thousand, a million Eves.
They are hard at work. They are dreaming
of all their daughters.

CLAYMAN

I

Joy of Clayman on a rainy day!
He sings loudly in the long-lived woodland.
He is no newcomer.
The hard-souled earth softens for him.
Rain falls floweringly on ferns and the bronze of brambles.
Kneeling, he digs coaxing hands into fertile earth.
She is his first love.
He inhales the equine sweat of her herbs,
the citrus-tendrilled blue lily,
the tree-stump-fungus' smoulder of myrrh,
the meditative stink of wild garlic, the drift and quiver
of wet leaf;
the boutonnière of earth is a sweet breath to him,
an inheritance for an impeccable son.
He moulds the earth in his fists. It draws his passion out.
He lugs the basket of ripe cloud-coloured clay home on his back,
unstumbling through the wet foliage.

Clayman has strong flexible hands,
earth on his hands has blessed and tended them,
his tender thumb swaddles and blurs the clay into shape,
is a tool of dominant comfort.
His blunt fingers hew, caress, knead and gouge.
The bachelor strokes out big-bellied bowls
and tall immaculate jars; they are unadorned,
he lifts them from earth only as far as he must.
Their memory of earth stills and satisfies them.

2

Clayman dreams of women.
Women are not still like these quieted bowls and silent jars.
Women move like trees spurring in autumn gales,
or like tides storming inland on fatherless wings.
They are all electricity, all shocks and brightness,
their bodies are light skirts shunting and dancing,
they are long-tongued quarrellers,
their voices high percussions, bee-in-a-bottle drawls.
But he dreams of their beautiful innocent lips.

He stares at his hands slabbed with clay.
The reveries and revenues of clay are his;
the white clay has the power a sleeping woman has,
her dreams scented with forests and fields.
The shape of a bowl is the woman shape,
he says, it is a womb holding water, it carries wine,
grain, milk, contains bone-ash of the dead, ultimate vessel.

Drinking from his clay cup, he drinks the past,
a wine of caprice.
Hauling open the kiln door to fire a new batch,
he relishes the roast heat, his man's oven,
that neat and purposeful hell.

My vessels return kindness for kindness, he says.
Touch for touch, earth kindles me into who I wish to be.

3

He puts the broken slabs of dry tough white clay
in an old tin bath and pees on it gently
to moisten it; the best way, the old way.
Then he leaves the clay to soak and soften.

Returning, he pinches the clay, testing its pliancy.
He puts his fingers to his mouth, tasting.
The sweet odd taste like wine and raw egg and mashed swede,
its smell of mould and rain and alcohol
tell him it's ready. He adds
a fistful of ground flint, to increase the whiteness,
a pinch of calcined animal bone, to add translucence
and blessing.

He lifts the clay in both hands
and thuds it down on the wooden benchtop,
knocking it into preliminary shape
by hammering it repeatedly with his fists,
then pressing the weight of his spread hands
down on it; the air must be forced out.
He grabs the clay up, throws it down,
beats it with his fists again. He punches
and pummels it, groaning and urging himself on;
it must be done;
this is not the gentle time.

With a wire he splices the clay in two, like cheese;
examines it for air bubbles.
Walloping the two halves together with a clap of laughter,
he wedges the clay, pushing the softest clay out
in convexing folds further than the firmest seams of clay.
After much adroit pushing and pulling with his hands,
much gripping and slapping, thwacking and thumping,
thrashing and pounding, the boneless clay is ready.

4

He holds the first slab of her,
preparing to create his love from clay,
(from the female genre of mud, earth,
clay, white silt).

He looks for the first hint of her
within the mass that he belabours,
as he moulds, as he finger-carves,
as he pulls off the many veils of clay that enwrap her;
in his hands the clay is shuddering;
loiteringly, he coils, kneads, shapes, coaxes.

Sweating, he plumps the mound of white clay
to a rough torso shape; a woman;
he hugs the clay to his own breast
as he works, shaping base of throat,
rib-cage, collar-bone, breasts, waist
and slight flare of hips; the white torso
presses hard against his flat hand;
they are breast to breast.
Under his hands he feels clay, gristle, sinew.
He sets her on the table.
With fresh clay, he rolls the cylinder that will be her arms.

5

He holds a damp white boulder of clay in his hands.
It is bigger than a woman's head,
for the clay will shrink as it dries.
He places the lump of clay gently on the wooden board.
It is white and lopsided, vague, blind.
He sits a little way off, waiting to recognize her;
waiting for the clay to give him permission
to look for the woman it wants to be.
He is scared; his skills, his conviction,
his husbandry fade. His dream of making a woman
out of clay, her voice instructing him fade.

The palms of his hands flutter with energy that eludes his grip;
he dare not turn to look at the finished torso under her cloak
of sacking, nor at the perfect arms and legs lying
under similar blankets; waiting.

Why am I making this woman?
I have made her a beautiful room,
with polished floor, mirrors, flowers,
with silent curtains, a bed of fluffy satin,
raisins and chicory to eat.
Why am I making her?

Not for a sister, not for a mother.
A companion? A friend?

Like an islander alone, he needs her.
She alone can explain why he needs her.

6

Now his hands brush the clay delicately,
gladly. He strokes,
he taps, he dabs. His smoothing hands
rock the gnarled clay. He shapes and turns and models,
he searches for her, he sings to her
in a light Italianate tenor, he unearths her,
he dis-covers her in the dollop of clay.
With a long knitting needle he sketches rough features,
then with fingers and brushes and knives retrieves them
into beauty; he looks for her, for the gift of her;
finding her in the element of clay not with his mechanic
fingers, his muscled arms or any tool,
but by the power of the atmospheres (like storm
and calm at once, all seasons) streaming through him to her,
bringing life closer and closer to her;
electrifyingly—
as he makes mouth, eyes, nostrils, ear-lobes
he murmurs to her so that she will not be afraid of these changes;

unhurriedly he knuckles out her eye-sockets,
he taps her face all over with a stiff brush, to contour
the surface; scouring, abrading, and then,
with the pads of his hands, patting the skin-softness in;
he wets his fingers, to pull her lips into shape,
not giving her a smile,
but letting her expression form naturally,
his fingers following the clay;
finding that she is fierce and sedate,
that her face reflects hope, anticipation, doubt, longing—
mirroring him,
yet drawing him towards a deeper debate;
what now?

He lifts her head.
He looks into her unseeing eyes.
He touches her lips with his own lips.
Holding his breath, he places her head upon the long stalk
of the waiting neck.
The clay woman stands as tall as Clayman.

He has shaped her lovingly, shoulders and spine,
neck and feet; in homage to her
he has given her a wide mouth, stately hips,
a flat belly, a deep idiomatic sex.
He has fashioned her as perfectly as he can,
using all his long-learned ways.
He has fingered out muscles, tendons, nerves,
remembering the bodies of women, their faces,
glances and frowns. But he has not copied from memory.
He made this woman of clay as if the woman stood whispering
directions in his ear, pointing with the sgraffito
what he must do or change, a more adventurous swagger
to the breasts, a stronger handspan, a milder
focus of shoulder.
He is the servant of her creation.
This is his strength and power, what makes the woman of clay
strong and powerful.

He dances around her to express his love;
he sings and pirouettes, leaping and tumbling in front of her;
but she is casual clay and stares like clay;
hears like clay.

      He kneels,
      lays his head against her belly.
      The clay is cool monotone,
      unchaperoning silence;
      her sex is just a groove in clay,
      her toes stiff, stubborn, nascent.

What now? What am I to do?
He gazes up at her; so perfect; lacking the gossamer
of breath, the peal of heart-beat.
He has made her as best he can;
the most perfect creature he could;
but he must go further. She must live.

7

It is night.
Each star burns with life.
He is naked. He stands in his skin before her.
All night he waits with her:
his body cooling, his limbs rigid with cold,
becoming like her. He feels the bliss
of being clay. He is true Clayman now.
Clayman and Claywoman are motionless.
This naked stillness of theirs marries them.
The stiff swathes of clay are their flesh and their finery.

8

At dawn she smiles at him.
He hears her voice trying first words.
She wades, steps, glides towards him.
She touches her eyelids, her lips.

51

Clayman is dumb. He is cold as stone.
Her body warms him. He trembles with cold and fear.
She is flesh and bold blood. Her body has plumes
and folds of heat. She takes him in her arms,
warming his cold clay back to life;
lets him weep and praise.
Her breath comes so easily, natural as all air.
No pain sews her together.
She is knit whole and complete, perfect
as a tree or a harp or a tear.
The first chalk-scrape of her voice
turns clear and tunable.
Her long willing hair, her warmth, her words shelter him.
He does not recall who created whom.
Love's transparent glaze slips over them.
They both began: now.

9

Claywoman wakes.
She tastes her own sweet saliva.
Claywoman chooses her own name.
Claywoman remembers the fine soft moon-smelling
earth, how she came from it, shaken through a sieve of sleep.
She remembers how he made her,
but that she is her mother's daughter.
She is named, reckless, self-indulgent, unpredictable,
affectionate, strong but soothing.
She breathes with delight,
sees everything for the first time,
light of day makeshift and lovely.
It belongs to her.
She hears rain falling, birds singing, people
talking in a babble. All is hers.
She will give it to him.
Her hand cupping her sex glows; she is rich;
she flexes her new limbs in fearless pleasure.
Her breasts scoop and shine.

She yawns and basks, kicking up her legs.
She watches her toes for a long time, loving them.
She smells her smell of sun-on-earth, of rain-on-leaf.
She combs her tangled hair.
Sleep has fired her in its kind kiln.

She smiles at Clayman opening her door.
She beckons him in, lovingly.
He kisses her with the patience of his craft.
He looks at her without complaint.
He wants to do everything for her.
So she shooes him away to fetch her breakfast.
'When that's done', she teases,
'you must sit and sew a new dress for me
with your big clumsy hands, oh your poor pricked fingers,
but you must do it before I will let you touch me,
even my little toe.'

He goes to the kitchen, happy as a person being taught love.
All his bowls and jugs have a meaning now.
His work is real to him now. He boils water for an egg,
toasts bread with passion.

Later, dressed in his best shirt,
Claywoman goes to the fields to visit her dear brothers and sisters,
saying,
'Last night I dreamt of when I was earth.
Now I am blood and milk. Now breath warms, scorches me.
Now I am changed and still changing,
but I will not forget my life as dirt, root, rot and worm.
Do not forget me.'

LEATHERMAN

I

He stands naked in his own skin
like water almost too deep to stand up in.
He is fond of his husband's shoulders,
his lover's haunches, his friend's feet.
In an interview with his own skin
he hears from his own lips
the intimate hint
that skin is a fine and blessèd linen,
yes, his skin, that another craftsman
might use, hang, stretch, tan and emboss,
set in the splat of a baronial chair
strong, lordly and chivalrous,
or soften so tenderly that a child
could wear his skin for her first shoes.
Oh the terror of skin for Leatherman!
Oh the velvet quiver of his peeled scalp,
the pang of his torn-out tongue!
How he howls in pain at agonies of slice and slash,
wheezing and choking
as if his skin sat flayed at his side,
companion-brother in cascades and lachrymaes
of blood-flow.
In horror he tastes and smells
his fountaining pooling blood. . . .
Oh the blood-rush ache in him for his wife,
who comes soothingly, mockingly, magically
when he bawls her name, 'Eva, oh Eva!'

Caressingly, light-heartedly,
she covers his fiery skin
with a robe of cool green-stem silk
she's sewn to protect him,
his poor harpooned skin,
his frightened pelt;
she takes him far out to sea on the breeze of her breath,

floating him away on the raft of her breast,
on the barge of her thighs;
he sails to safety on the ship
of her broad herculean pelvis, the dark sails
of her skin taut, sturdy and felicitous,
surging and redeeming round him.

2

Leatherman is brave today.
Stout and bearded, bull-like,
he is beast-owner, tamer of hides.
His work begins with death.
Cattle die for him,
they let their hides soften and rot
and become items of luxury in his hands.
His hands are soft from the lanolin
and grease he rubs into the leather,
shining it like a summer bargain,
saddles, belts, bags, skirts, jackets,
elaborate and costly objects he makes with love,
hacking and slashing and scissoring,
then delicately annexing seam to seam,
fashioning the leather blamelessly.

His workshop has the rainy smell of a stable,
leather's shock whiff of dung and leaf,
undiluted by the mild aloof perfumes of the arcades
and precincts which will soon display his goods;
Leatherman is drunk on this first rough odour,
and then he frowns at praise,
diving deep into the balsams and resins of leather,
gripping his tough needles, his webs of strong thread.

3

Leatherman has a wife whose skin is black.
She is his black joy.
Her black-skinned body is a gift that she gives and gives.
Oh he is a soft-pedaller, she knows!
He is no stronger than his hands.
He is the groom of her skin.
Her blackness is a deer in his heart,
a salve on his own pale meaningless skin.
He shelters from his fears with her.
He makes her shoes and skirts
of wistful white leather that revere her blackness.
He makes trousers for her of leather black as her skin,
that grip her black thighs with an erotic piety
she allows him to have; he makes her
a leather coat of unimpeachable sinfulness;
he makes her a pair of suede gloves the colour of blood,
a perfect fit, light as an ounce of red evenings.

4

Now he has made his wife a bed of supple leather
for her to lie and fuck with him,
her black cunt strong and shrewd round his sweet cock,
straddling him, raising the dead,
rising and falling, gasping, puffing,
her breasts huge and innocent above him, her nipples
long, straight, brown — and a smell of blood as she comes,
groaning, far-gone; and with a lift and leap of his hips,
he floods into her, docile and weeping.
She looks down at him, fanning out her long slender toes.
He is her lover of leather.
Her sleep on their leather bed is bold, witty and naked.

5

For the nights of summer
Leatherman has made his wife
a billowing tent of skins,
a shed of leather hugging the ground,
where the sky cannot come.

Here with the thin whistling wands of charity
the wife of Leatherman beats him like one of his own hides,
pegged out at her loving feet.
His howls are the song of love as it burns into flesh.
Rods of rose and henna ray out across the white of his loins.
His skin is what he wears without secrecy.
He needs touch and triple touch, he needs her accuracy,
the sting and strafe of pain, this is the life of the skin,
this is his way in,
he is leather and suffers like it;
he glows like a god, one of the bruised angels from hell.

GLASSMAN

Glassman is thin and quick and sentimental.
His shaded eyes spill with tears. He is lonely.
How young he is. He dreamed of a glass labyrinth.
When he woke he called it, 'My Moon'.
In his studio Glassman is like a boy at conspiratorial play,
making things.

He has knowledge of orient glass,
of the glittering pregnancies of transparent goblets.
He has animals of glass. He has birds clear as silence.
Humming-birds, owls, swans!
He has linnets and pigeons of glass. He has gardens
of glass. Flowers, petals clearly sheathed in glass leaves.
He has apples, lilies and lilacs of glass.
He has pears and pineapples, cabbages and caviare of glass.
He has rooms of glass angles, disappearance rooms,
the humour of glass as it vanishes. He has a horse
of galloping glass. He has dancers,
he has fish of glass, his clear swimmers,
he has a silent mandolin of glass, a complacent shell,
a galleon bearded with glass sails.

2

Oh the laverock dart of the glass as
he blows into its heat! His delicate favourable jaw.
His loving lips flicker and smile. He blows again,
gentle huffs and puffs into fiery glass
that swells and coils. With a sound of slapped water
his breath goes smooth into glass,
bribing and bubbling it into shape.
He blows mood and meaning into each morsel and gobbet
of glass.
His young throat tenses, cooling.
His blue eyes mimic all blues.

3

Glassman wants his birds of glass to sing.
He wants his lizards and snakes to creep and crawl.
He longs to ride his glass horse,
to taste his glass fruit, swim with his fish of glass.
He needs to smell his flowers of glass, snowdrop and sunflower.
But transparence keeps them still and silent.
He cannot hear them, taste them, smell them, brother them.
His creatures have no memory.
His breath forged them. They swooped into silent being,
motionless, waiting.
He asks glass to breathe like him.
He asks glass to live,
his words calm as the great sighs of dogs
resting. Young, stern, unembarrassed,
he spoke to his creations about life.

4

Glass did not come to life. The birds, the flowers,
the snakes stayed glass. Maybe glass was not ready to live
like flesh or petal. Maybe glass was happy as glass.

Glassman did not ask again.
He slept deeply, a man whose work can be seen through.

Morning daylight blue and bare ran over his glass kin
with such love that he woke to gaze for hours at their splendour.

5

Glassman makes a woman of sweet glass.
She has the assumed innocence of water.
She has poverty, chastity and obedience.
She is blown from his fiery dreaming sperm,
she comes from the scald of his desire,
bringing him good luck.
Her horoscope of glass promises fortune.

Her heart of glass beats slowly.
Her breath is a long drift of glass.
Her glass nose-bone is individual and perfect.
Her embrace is cool and delicate as any monarch.

She watches him build her a house of glass,
a flashing shack of glass, hot in sun, cool in moon,
all as she orders,
her roof flittering and scorching,
her walls all window, her windows all wall: glass.
In this house he sweats and worships,
Adam and his see-through Eve.

For her he succumbs to glass.
He is true Glassman now, her consort.
His flesh slouches off him in shards.
He is glass. His beard of glass glows,
his eyes of glass splash glass tears of bliss,
his prick of glass stiff-dangles,
his glass arms reach out to her.

In her nakedness of glass
she comes towards him carrying a bird of glass.
Husband and wife of glass sigh as it flutters wings of glass,
sun-lit throat swollen and trembling and pulsing.
She throws the glass bird up into the sky of glass,
tossing it high; it rises alive and singing
into the open morning; into the slant of blue.

Now he and she may settle down in their Eden,
where glass can share the fates of flesh,
where flesh is as true a spirit as the clear broad glass.

WIVES AND WORKERS

These are the three workers, with their materials,
clay, leather, glass;
here are their wives, Lady Glass, Lady Black-Leather, and Lady
　　　Clay.

The men work in daylight,
they have the hands of their fathers,
the eyes and hearts of their mothers.

Clayman labours with the soft solid burden of clay.
He loves it like the ground under his feet, with ordinary
useful love. He develops his husbandry,
his hands unfaltering in the daylong odour of earth.
Above his head, the roof shines with gold
from ridge to eaves, the gilded house of the married man.

Leatherman unreels a long length of indolent tanned skin,
jetsam delicate as air. He strokes its spiced weave
like pulled-out lace from a lavish corsetry.
His blood-roots quicken and quiver like a beaten drum of feathers.
He works fast with template and broad knife; giving
leather its after-life.
It is his gift, flagrant, simple and secret.

Glassman breathes in a wisp of carbons.
Torment of glass dust hides in the mist of his lungs.
His lips murmur; incantation and joke,
incandescent forfeit.
He makes a crescent moon of wild robust glass,
letting the glass flow and shift, adaptable,
the vernacular of glass rustling.

The wives watch.

One came from the willing earth,
to live in his helpless boudoir,
under his bright sloping roof;
so that he will never be homesick.

One came from her man's love of leather;
her love is thin as sky, tough as whipcord.
His heart hangs from her wide and studded leather belt.

One came from his glass-wet dream,
his potent sigh;
her love is his one strong plank
from a wrecked ship of glass.
Over hurdles of glass, she leaps towards him.

One loves Glass and marries him.
One loves leather and marries Leather.
One loves earth and marries Clay.
There are long silent busy nuptials
in the nights;
each woman burns her own tongue in the flame of love.

Each Eva permits an Adam to create her;
so she may begin her work,
which is to finish his work for him.

Claywoman dissolves the pots and bowls in rain,
bringing them down to earth again.

Leatherwoman lets the killed leather graze,
solid from horn to hoof. The herds
look up, surprised, then feed again, big tongues
simplifying grass.

Glasswoman sets glass singing; the birds flock
and fuss; next, fruit ripens into colour, fragrance,
eloquent seed; real fish grunt through real water;
a world of glass goes free, learns life,
the weight of death.

The Eves are here. The Eves are dancing
and working, whispering and flattering, deceiving
and observing, crouching and licking, stretching
and bargaining; the Eves are here,
a hundred, a thousand, a million Eves.
They are hard at work. They are dreaming of all their daughters.

OXFORD POETS

Fleur Adcock

James Berry

Edward Kamau Brathwaite

Joseph Brodsky

Michael Donaghy

D. J. Enright

Roy Fisher

David Gascoyne

David Harsent

Anthony Hecht

Zbigniew Herbert

Thomas Kinsella

Brad Leithauser

Herbert Lomas

Derek Mahon

Medbh McGuckian

James Merrill

John Montague

Peter Porter

Craig Raine

Tom Rawling

Christopher Reid

Stephen Romer

Carole Satyamurti

Peter Scupham

Penelope Shuttle

Louis Simpson

Anne Stevenson

George Szirtes

Anthony Thwaite

Charles Tomlinson

Chris Wallace-Crabbe

Hugo Williams

also

Basil Bunting

Keith Douglas

Ivor Gurney

Edward Thomas